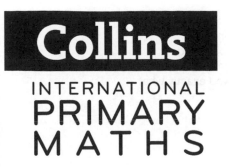

Collins

INTERNATIONAL PRIMARY MATHS

Progress Book 2

William Collins' dream of knowledge for all began with the publication of his first book in 1819. A self-educated mill worker, he not only enriched millions of lives, but also founded a flourishing publishing house. Today, staying true to this spirit, Collins books are packed with inspiration, innovation and practical expertise. They place you at the centre of a world of possibility and give you exactly what you need to explore it.

Collins. Freedom to teach.

Published by Collins
An imprint of
HarperCollins*Publishers*
The News Building
1 London Bridge Street
London
SE1 9GF

HarperCollins*Publishers*
Macken House,
39/40 Mayor Street Upper,
Dublin 1, D01 C9W8, Ireland

Browse the complete Collins catalogue at
www.collins.co.uk

© HarperCollins*Publishers* Limited 2021

10 9 8 7

ISBN 978-0-00-836958-3

British Library Cataloguing-in-Publication Data
A catalogue record for this publication is available from the British Library.

Author: Peter Clarke
Series editor: Peter Clarke
Publisher: Elaine Higgleton
Product developer: Holly Woolnough
Copyeditor: Tanya Solomons
Proofreader: Catherine Dakin
Answer checker: Steven Matchett
Cover designer: Gordon MacGilp
Cover illustrator: Ann Paganuzzi
Illustrator: Ann Paganuzzi
Typesetter: Ken Vail Graphic Design Ltd
Production controller: Lyndsey Rogers
Printed and bound in the UK using 100% Renewable Electricity at CPI Group (UK) Ltd

MIX
Paper | Supporting
responsible forestry
FSC www.fsc.org   FSC™ C007454

This book is produced from independently certified FSC™ paper to ensure responsible forest management.

For more information visit: www.harpercollins.co.uk/green

Photo acknowledgements
Every effort has been made to trace copyright holders. Any omission will be rectified at the first opportunity.
p12t Tartila/Shutterstock; p12b Fotosv/Shutterstock; p14b Darsi/Shutterstock; p24 Mictoon/Shutterstock; p26t Hit Toon/Shutterstock; p26b Insdesign86/Shutterstock; p27t Moving Moment/Shutterstock; p27c Ron Dale/Shutterstock; p27b Antpkr/Shutterstock; p28t MsVector Plus/Shutterstock; p28b KittyVector/Shutterstock; p47 Szefei/Shutterstock; p49tr Flowersmile/Shutterstock; p49c HappyPictures/Shutterstock; p49bl Tanadtha Iomakul/Shutterstock; p149br Stanislav Stradnic/Shutterstock; p50tl GraphicsRF.com/Shutterstock; p50tcl Kindlena/Shutterstock; p50tc Svetlana Guteneva/Shutterstock; p50tcr Pingebat/Shutterstock; p50tr Stockakia/Shutterstock; pp52-3 Ghenadie/Shutterstock; p54tl Benchart/Shutterstock; p57tl Lady-luck/Shutterstock; p57c Brgfx/Shutterstock; p58c Duda Vasilii/Shutterstock; p58b BudOlga/Shutterstock; p59tl Totallyjamie/Shutterstock; p59tc VectorSun/Shutterstock; p59tr BlueRingMedia/Shutterstock; p59cl ALMAGAMI/Shutterstock; p59cr BlueRingMedia/Shutterstock; p59bl PhotoStockImage/Shutterstock; p59bcl Alekseykolotvin/Shutterstock; p59bcr Terdpong/Shutterstock; p59br Irbena/Shutterstock; p59 Zonda/Shutterstock; p60t KajaNi/Shutterstock; p60c Bannykh Alexey Vladimirovich/Shutterstock; p60b Spreadthesign/Shutterstock; p63 4zevar/Shutterstock.

The publishers gratefully acknowledge the permission granted to reproduce the copyright material in this book. Every effort has been made to trace copyright holders and to obtain their permission for the use of copyright material. The publishers will gladly receive any information enabling them to rectify any error or omission at the first opportunity.

Cambridge International copyright material in this publication is reproduced under licence and remains the intellectual property of Cambridge Assessment International Education

This text has not been through the Cambridge International endorsement process.

Contents

Introduction   4

I can statements   5–9

Number

1   Counting and sequences (A)   10
2   Counting and sequences (B)   12
3   Reading and writing numbers to 100   14
4   Addition and subtraction   16
5   Addition   18
6   Subtraction   20
7   Multiplication as repeated addition   22
8   Multiplication as an array   24
9   Division as sharing   26
10   Division as grouping   28
11   Division as repeated subtraction   30
12   Times tables (A)   32
13   Times tables (B)   34
14   Money   36
15   Place value and ordering   38
16   Place value, ordering and rounding   40
17   Fractions (A)   42
18   Fractions (B)   44

Geometry and Measure

19   Time   46
20   2D shapes, symmetry and angles   48
21   3D shapes   50
22   Length   52
23   Mass   54
24   Capacity and temperature   56
25   Position, direction, movement and reflection   58

Statistics and Probability

26   Statistics   60
27   Statistics and chance   62

The Thinking and Working Mathematically Star   64

Introduction

The Progress Books include photocopiable end-of-unit progress tests which are designed to assist teachers with medium-term 'formative' assessment.

Each test is designed to be used within the classroom at the end of a Collins International Primary Maths unit to help measure the progress of learners and identify strengths and weaknesses.

Analysis of the results of the tests helps teachers provide feedback to individual learners on their specific strengths and areas that require improvement, as well as analyse the strengths and weaknesses of the class as a whole.

Self-assessment is also an important feature of the Progress Books as feedback should not only come from the teacher. The Progress Books provide opportunities at the end of each test for learners to self-assess their understanding of the unit, as well as space for teacher feedback.

Structure of the Progress Books

There is one progress test for each of the 27 units in Stage 2.

Each test consists of two pages of questions aimed at assessing the learning objectives from the Cambridge Primary Mathematics Curriculum Framework (0096) for the relevant unit. Where appropriate, this also includes questions to assess learners' development in one or more of the Thinking and Working Mathematically characteristics, as indicated by the TWM star on the page. All of the questions are typeset on triangles to indicate that they are suitable for the majority of learners and assess the unit's learning objectives.

Pages 5 to 9 include a list of photocopiable *I can* statements for each unit which are aimed at providing an opportunity for learners to undertake some form of self-assessment. The intention is that once learners have answered the two pages of questions, they turn to the *I can* statements for the relevant unit and think about each statement and how easy or hard they find the topic. For each statement they colour in the face that is closest to how they feel:

☺ I can do this   ☺ I'm getting there   ☹ I need some help.

A photocopiable variation of the Thinking and Working Mathematically Star is also included in the Progress Books. This version of the star includes *I can* statements for the eight TWM characteristics. Its purpose is to provide an opportunity for learners, twice a term/semester, to think about each of the statements and record how confident they feel about Thinking and Working Mathematically.

Administering each end-of-unit progress test

Recommended timing: 20 to 30 minutes, although this can be altered to suit the needs of individual learners and classes.

Before starting each end-of-unit progress test, ensure that each learner has the resources needed to complete the test. If needed, resources are listed in the 'You will need' box at the start of each test.

On completion of each end-of-unit progress test, use the answers and mark scheme available as a digital download to mark the tests.

Use the box at the bottom of the second page of each end-of-unit progress test to either:

- write the number of marks achieved by the learner out of the total marks possible.

- sign or initial your name to indicate you have marked the test.

- draw a simple picture or diagram such as one of the three faces (☺, ☺, ☹) to indicate your judgement on the learner's level of understanding of the unit's learning objectives.

- write a brief comment such as 'Well done!', 'You've got it', 'Getting there' or 'See me'.

Provide feedback to individual learners as necessary on their strengths and the areas that require improvement. Use the 'Class record-keeping document' located at the back of the Teacher's Guide and as a digital download to update your judgement of each learner's level of mastery in the relevant sub-strand.

I can statements

At the end of each unit, think about each of the *I can* statements and how easy or hard you find the topic. For each statement, colour in the face that is closest to how you feel.

Unit 1 – Counting and sequences (A)	Date:			
• I can count up to 100 objects.		☺	😐	☹
• I can recognise how many objects there are without counting them.		☺	😐	☹
• I can count on and back in fives to 100.		☺	😐	☹
• I can count on and back in tens to 100.		☺	😐	☹
Unit 2 – Counting and sequences (B)	**Date:**			
• I can count up to 100 objects.		☺	😐	☹
• I can count on and back in steps of 1, 2, 5 and 10 to 100.		☺	😐	☹
• I can recognise odd and even numbers to 100.		☺	😐	☹
• I can estimate the number of objects to 100.		☺	😐	☹
Unit 3 – Reading and writing numbers to 100	**Date:**			
• I can count on in ones to 100.		☺	😐	☹
• I can count back in ones from 100.		☺	😐	☹
• I can read the numbers and number names to 100.		☺	😐	☹
• I can write the numbers and number names to 100.		☺	😐	☹
Unit 4 – Addition and subtraction	**Date:**			
• I understand and can explain the relationship between addition and subtraction.		☺	😐	☹
• I know pairs of numbers that total 20.		☺	😐	☹
• I can answer number sentences, such as 50 + 40 and 80 – 30.		☺	😐	☹
• I can add more than two numbers together.		☺	😐	☹

Unit 5 – Addition		Date:			
• I can add a 1-digit number to a 2-digit number, such as 46 + 3.			☺	😐	☹
• I can add a tens number to a 2-digit number, such as 28 + 40.			☺	😐	☹
• I can add two 2-digit numbers, such as 34 + 23.			☺	😐	☹
• I can estimate the answer to an addition.			☺	😐	☹
Unit 6 – Subtraction		Date:			
• I can subtract a 1-digit number from a 2-digit number, such as 67 – 5.			☺	😐	☹
• I can subtract a tens number from a 2-digit number, such as 58 – 30.			☺	😐	☹
• I can subtract two 2-digit numbers, such as 76 – 32.			☺	😐	☹
• I can estimate the answer to a subtraction.			☺	😐	☹
Unit 7 – Multiplication as repeated addition		Date:			
• I can count objects in equal groups to solve a multiplication.			☺	😐	☹
• I can draw a picture or use a number line to multiply.			☺	😐	☹
• I can use × and = to write a multiplication number sentence.			☺	😐	☹
Unit 8 – Multiplication as an array		Date:			
• I understand multiplication as an array.			☺	😐	☹
• I can draw an array to multiply.			☺	😐	☹
• I understand that an array shows two multiplications with the same answer.			☺	😐	☹
Unit 9 – Division as sharing		Date:			
• I can share objects equally between a given number of sets to find out how many are in each set.			☺	😐	☹
• I can use ÷ and = to write a division number sentence.			☺	😐	☹

		☺	😐	☹
Unit 10 – Division as grouping	Date:			
• I can put objects into groups of equal size to find out how many groups there are.		☺	😐	☹
Unit 11 – Division as repeated subtraction	Date:			
• I can solve a division by subtracting groups of the same size and counting how many groups there are.		☺	😐	☹
Unit 12 – Times tables (A)	Date:			
• I know the multiplication facts for the 2 times table.		☺	😐	☹
• I know the division facts for the 2 times table.		☺	😐	☹
• I know the multiplication facts for the 5 times table.		☺	😐	☹
• I know the division facts for the 5 times table.		☺	😐	☹
Unit 13 – Times tables (B)	Date:			
• I know the multiplication facts for the 10 times table.		☺	😐	☹
• I know the division facts for the 10 times table.		☺	😐	☹
• I know the multiplication facts for the 1 times table.		☺	😐	☹
• I know the division facts for the 1 times table.		☺	😐	☹
Unit 14 – Money	Date:			
• I can recognise the symbol we use for money.		☺	😐	☹
• I can compare values of coins and notes.		☺	😐	☹
• I can find coins and notes that have the same value.		☺	😐	☹
Unit 15 – Place value and ordering	Date:			
• I can say the number of tens and ones in a number to 100.		☺	😐	☹
• I can compare two numbers to 100.		☺	😐	☹
• I can use ordinal numbers to show position.		☺	😐	☹

Unit 16 – Place value, ordering and rounding	Date:			
• I can join groups or sets of numbers to make another number, such as 50 + 7 = 57.		☺	😐	☹
• I can break a number into tens and ones, such as 32 = 30 + 2.		☺	😐	☹
• I can break up a number in different ways, such as 32 = 10 + 10 + 10 + 2.		☺	😐	☹
• I can compare and order numbers to 100.		☺	😐	☹
• I can round a number to 100 to the nearest 10.		☺	😐	☹
Unit 17 – Fractions (A)	Date:			
• I can recognise objects and shapes that are in quarters.		☺	😐	☹
• I can find one quarter of an object or a shape.		☺	😐	☹
• I can find one quarter of a set of objects.		☺	😐	☹
Unit 18 – Fractions (B)	Date:			
• I can recognise different fractions that are the same amount.		☺	😐	☹
• I can combine halves and quarters to make other fractions.		☺	😐	☹
• I understand fractions as division.		☺	😐	☹
Unit 19 – Time	Date:			
• I can use and read a calendar.		☺	😐	☹
• I can order time.		☺	😐	☹
• I can read and show times to five minutes.		☺	😐	☹
Unit 20 – 2D shapes, symmetry and angles	Date:			
• I can recognise, describe and sort 2D shapes.		☺	😐	☹
• I can recognise horizontal and vertical lines of symmetry.		☺	😐	☹
• I understand that an angle is a measure of a turn.		☺	😐	☹

		☺	😐	☹
Unit 21 – 3D shapes	Date:			
• I can recognise, describe and sort 3D shapes.		☺	😐	☹
Unit 22 – Length	Date:			
• I can estimate and measure lengths in centimetres and metres.		☺	😐	☹
• I can use a ruler to draw and measure lines.		☺	😐	☹
Unit 23 – Mass	Date:			
• I can estimate and measure mass in grams and kilograms.		☺	😐	☹
• I can use scales to measure and compare mass.		☺	😐	☹
Unit 24 – Capacity and temperature	Date:			
• I can estimate and measure capacity in litres and millilitres.		☺	😐	☹
• I can use a thermometer to read, interpret and compare temperature.		☺	😐	☹
Unit 25 – Position, direction, movement and reflection	Date:			
• I can sketch the reflection of a 2D shape.		☺	😐	☹
• I can give and follow directions.		☺	😐	☹
• I can recognise whole, half and quarter turns.		☺	😐	☹
Unit 26 – Statistics	Date:			
• I can use a tally chart to collect and interpret data.		☺	😐	☹
• I can use a pictogram to present data.		☺	😐	☹
• I can use a block graph to present data.		☺	😐	☹
• I can use a Venn or Carroll diagram to sort data.		☺	😐	☹
Unit 27 – Statistics and chance	Date:			
• I can recognise and describe patterns.		☺	😐	☹
• I can describe the chance of something happening.		☺	😐	☹
• I can investigate chance and record the results.		☺	😐	☹
• I can present and describe data in a chart or graph.		☺	😐	☹

Number

Name: _____

1 Count the pearls. ☐

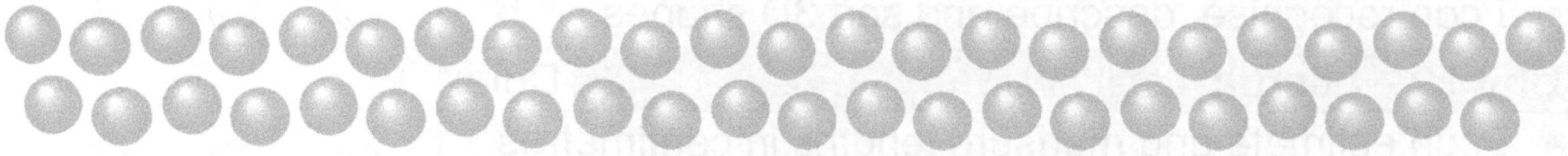

2 Count the beads. ☐

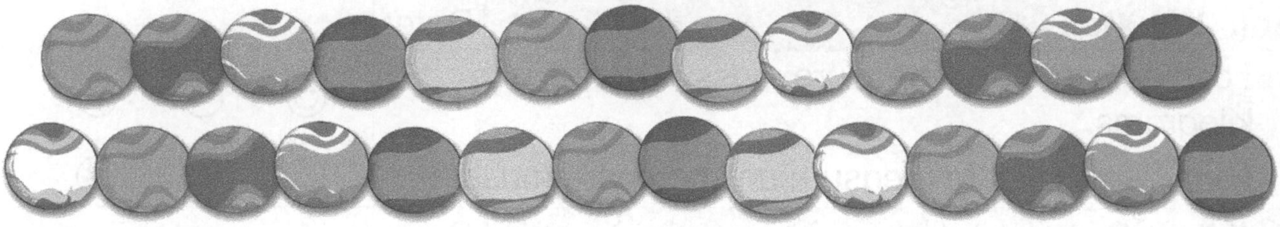

3 Without counting, write how many dots there are in each ten frame.

a ☐

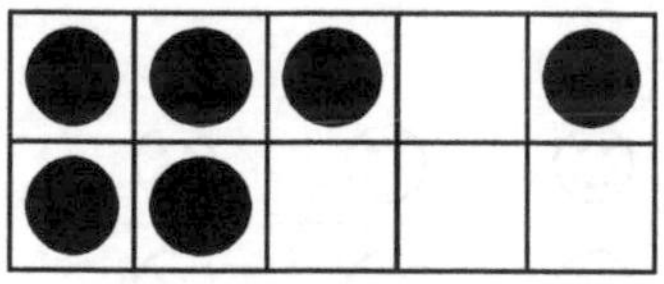

b ☐

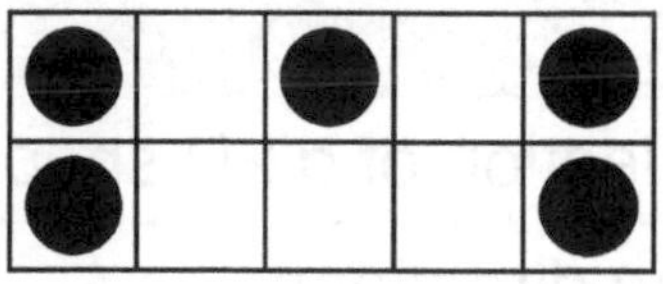

4

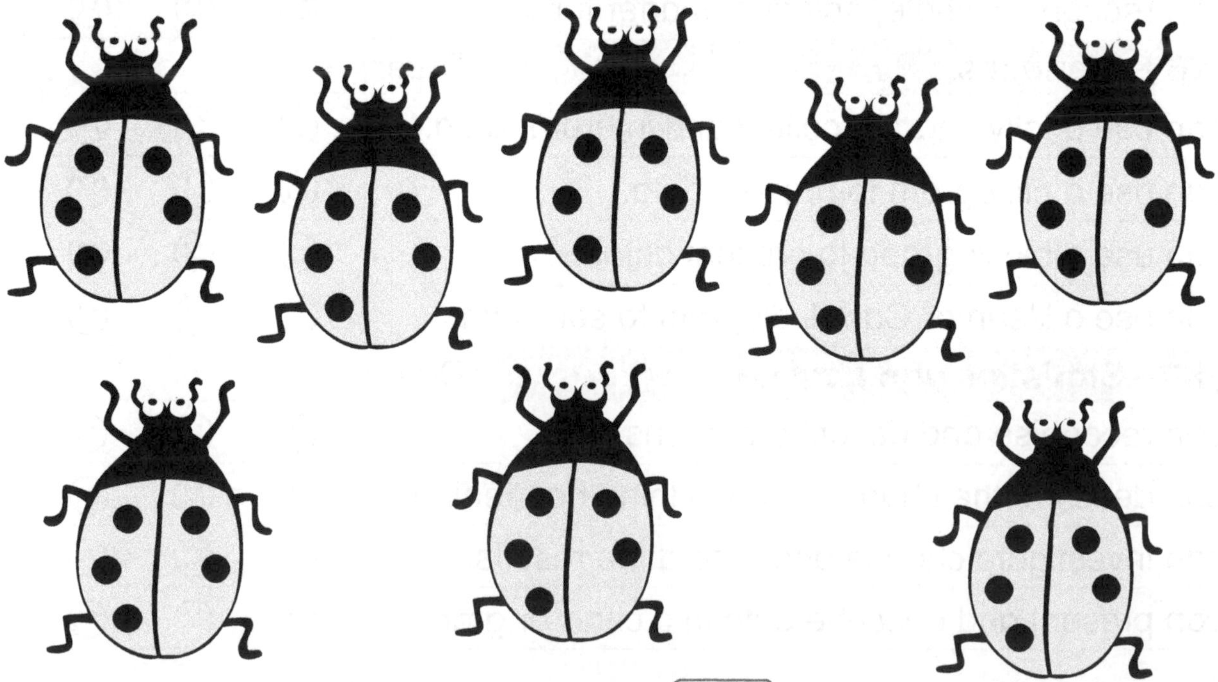

a How many ladybirds are there? ☐

b How many spots altogether? ☐

5 Write the next number in each sequence.

a 10, 15, 20, 25, ⬚   **b** 35, 40, 45, 50, ⬚

c 90, 85, 80, 75, ⬚   **d** 65, 60, 55, 50, ⬚

6

a How many butterflies are there? ⬚

b How many spots altogether? ⬚

7 Write the next number in each sequence.

a 10, 20, 30, 40, ⬚   **b** 60, 50, 40, 30, ⬚

c 90, 80, 70, 60, ⬚   **d** 50, 60, 70, 80, ⬚

Now look at and think about each of the *I can* statements. ⬚

Date: _____

Number

Name: _____

1 Count the jewels. ☐

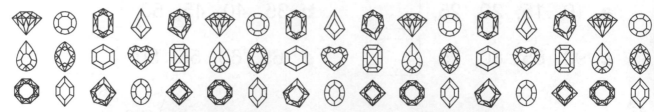

2

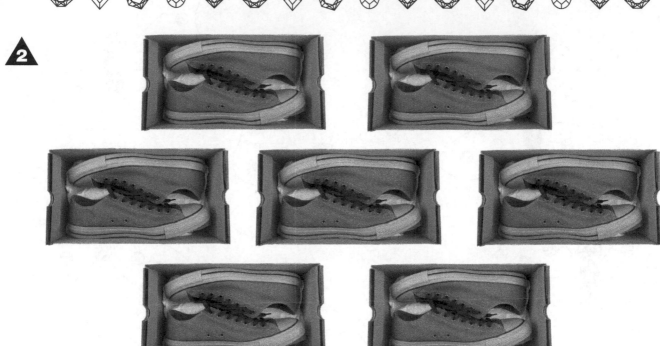

a How many boxes of shoes are there? ☐

b How many shoes altogether? ☐

3 a Circle all the **even** numbers.

23   42   60
  78   91

57   19   6
   85   34

b Write three other **even** numbers. ☐ ☐ ☐

c Write three other **odd** numbers. ☐ ☐ ☐

Number

 Continue the number patterns.

a 6, 8, 10, 12, ☐ , ☐ , ☐

b 80, 70, 60, 50, ☐ , ☐ , ☐

c 95, 90, 85, 80, ☐ , ☐ , ☐

d 38, 36, 34, 32, ☐ , ☐ , ☐

e 25, 30, 35, 40, ☐ , ☐ , ☐

f 30, 40, 50, 60, ☐ , ☐ , ☐

 Circle the estimate that you think is closest to the number of sweets in each jar.

a   20   40   **b**   20   40

      60   100   60   100

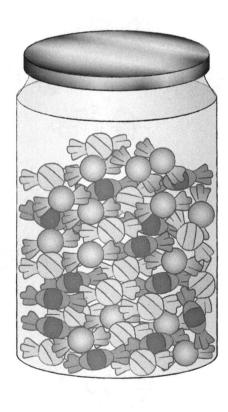

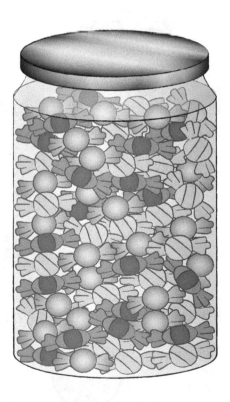

Now look at and think about each of the *I can* statements.

Date: _____

Name: _____

Number

1 Continue the line to count **on** from 64 to 83.

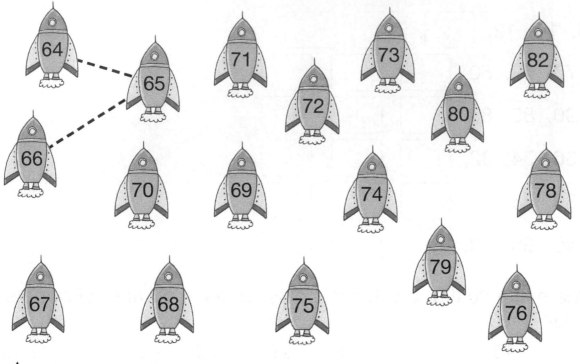

2 Continue the line to count **back** from 57 to 38.

3 Draw lines to match the numbers.

| 96 | 34 | 71 | 20 |

| 71 | 20 | 96 | 34 |

4 Fill in the missing numbers.

61	62		64	65		67	68		70
	72	73	74		76		78	79	
81		83		85	86	87		89	90

5 Draw lines to match the numbers to their number names.

62 •   • eighty-seven

35 •   • sixteen

87 •   • sixty-two

54 •   • thirty-five

16 •   • fifty-four

6 Write the matching numbers.

a twenty-one ☐   b seventy-three ☐

c ninety-eight ☐   d forty-nine ☐

7 Write the matching number names.

a 65 ☐   b 53 ☐

c 31 ☐   d 86 ☐

Now look at and think about each of the *I can* statements. ☐

Date: _____

Number

Name: _____

 Work out the answer. Then write the three matching number sentences.

a 8 + 4 = ☐

☐ + ☐ = ☐

☐ − ☐ = ☐

☐ − ☐ = ☐

b 7 + 3 = ☐

☐ + ☐ = ☐

☐ − ☐ = ☐

☐ − ☐ = ☐

 Complete the diagrams to show pairs of numbers that total 20. Then complete the number sentences.

a

20	
9	

9 + ☐ = 20

20 − ☐ = 9

☐ + 9 = 20

20 − 9 = ☐

b

20	
14	

14 + ☐ = 20

20 − ☐ = 14

☐ + 14 = 20

20 − 14 = ☐

 Write two different pairs of numbers that total 20.

☐ + ☐ = 20

☐ + ☐ = 20

Number

 Write an addition and subtraction fact for these tens.

a

60	
40	20

☐ + ☐ = ☐

☐ − ☐ = ☐

b

80	
50	30

☐ + ☐ = ☐

☐ − ☐ = ☐

 Complete each number sentence.

a 50 + 20 = ☐

b 40 − 10 = ☐

c 60 − 30 = ☐

d 20 + 30 = ☐

e 40 + 50 = ☐

f 90 − 30 = ☐

 Complete each number sentence.

a 2 + 5 + 3 = ☐

b 4 + 1 + 2 = ☐

c 1 + 2 + 3 = ☐

d 3 + 2 + 4 = ☐

e 2 + 2 + 3 + 1 = ☐

f 1 + 3 + 2 + 1 = ☐

Now look at and think about each of the *I can* statements.

Date: _____

Number

Name: _____

 Use the number lines to work out the answers.

a 45 + 4 = ☐

b 3 + 86 = ☐

c 5 + 72 = ☐

 Use the number lines to work out the answers.

a 42 + 30 = ☐

b 34 + 50 = ☐

c 51 + 40 = ☐

3 Use the number lines to work out the answers. Estimate first.

a 34 + 23 = ☐

Estimate: ☐

b 26 + 52 = ☐

Estimate: ☐

c 35 + 44 = ☐

Estimate: ☐

Number

 Use your own method to work out the answers.
Show your working out.

a 63 + 5 = ☐

b 29 + 50 = ☐

c 63 + 14 = ☐

d 41 + 26 = ☐

 Work out the answers using the expanded written method.
Estimate first.

a 53 + 25 = ☐

Estimate: ☐

b 44 + 13 = ☐

Estimate: ☐

c 62 + 27 = ☐

Estimate: ☐

Now look at and think about each of the *I can* statements. ☐

Date: _____

Number

Name: _____

1 Use the number lines to work out the answers.

a $67 - 3 =$ ⬚

b $88 - 5 =$ ⬚

c $79 - 4 =$ ⬚

2 Use the number lines to work out the answers.

a $76 - 40 =$ ⬚

b $91 - 60 =$ ⬚

c $53 - 20 =$ ⬚

3 Use the place value charts to work out the answers. Estimate first.

a $68 - 25 =$ ⬚

Estimate: ⬚

b $87 - 53 =$ ⬚

Estimate: ⬚

10s	1s
| | | | | |	○○○○○
|	○○○

10s	1s
| | | | | |	○○○○○
| | |	○○

4 Use your own method to work out the answers.
Show your working out.

a 55 − 4 = ☐

b 82 − 50 = ☐

c 59 − 24 = ☐

d 95 − 42 = ☐

5 Work out the answers using the expanded written method.
Estimate first.

a 59 − 31 = ☐

Estimate: ☐

b 75 − 44 = ☐

Estimate: ☐

c 47 − 23 = ☐

Estimate: ☐

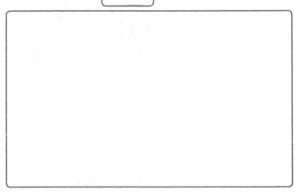

Now look at and think
about each of the
I can statements.

☐

Date: _____

Name: _____

Number

1 Count the cubes in their groups to solve the multiplications.

a 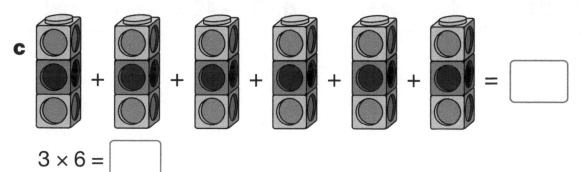 + + + = [ ]

$2 \times 4 =$ [ ]

b + + + + = [ ]

$4 \times 5 =$ [ ]

c + + + + + = [ ]

$3 \times 6 =$ [ ]

2 Draw crosses in the diagrams to solve the multiplications.

a $4 \times 6 =$ [ ]

✗ ✗ ✗ ✗					

b $6 \times 2 =$ [ ]

✗ ✗ ✗ ✗ ✗ ✗	

c $10 \times 3 =$ [ ]

✗ ✗ ✗ ✗ ✗ ✗ ✗ ✗ ✗ ✗		

d $5 \times 7 =$ [ ]

✗ ✗ ✗ ✗ ✗						

Number

3 Draw the correct number of jumps to match each multiplication. Then write the answer.

a 2 × 6 = [ ]

```
←─┬──┬──┬──┬──┬──┬──┬──┬──┬──┬──┬──┬──→
  0  1  2  3  4  5  6  7  8  9 10 11 12
```

b 3 × 3 = [ ]

```
←─┬──┬──┬──┬──┬──┬──┬──┬──┬──┬──┬──┬──→
  0  1  2  3  4  5  6  7  8  9 10 11 12
```

c 4 × 2 = [ ]

```
←─┬──┬──┬──┬──┬──┬──┬──┬──┬──┬──┬──┬──→
  0  1  2  3  4  5  6  7  8  9 10 11 12
```

4 Use the 100 square to help you to solve the multiplications.

a 5 × 3 = [ ]

b 10 × 6 = [ ]

c 2 × 8 = [ ]

d 5 × 9 = [ ]

e 10 × 9 = [ ]

1	2	3	4	5	6	7	8	9	10
11	12	13	14	15	16	17	18	19	20
21	22	23	24	25	26	27	28	29	30
31	32	33	34	35	36	37	38	39	40
41	42	43	44	45	46	47	48	49	50
51	52	53	54	55	56	57	58	59	60
61	62	63	64	65	66	67	68	69	70
71	72	73	74	75	76	77	78	79	80
81	82	83	84	85	86	87	88	89	90
91	92	93	94	95	96	97	98	99	100

Now look at and think about each of the *I can* statements. [ ]

Date: _____

Number

Name: _____

1 Write and solve a number sentence for each array.

a

[ ] × [ ] = [ ]

b

[ ] × [ ] = [ ]

c

[ ] × [ ] = [ ]

d

[ ] × [ ] = [ ]

2 Draw an array to match each number sentence.

a $2 \times 6 =$ [ ]

b $4 \times 4 =$ [ ]

c $3 \times 5 =$ [ ]

d $6 \times 3 =$ [ ]

3 Write an equal number statement to match each array.

a    ☐ × ☐ = ☐ × ☐

b 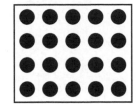   ☐ × ☐ = ☐ × ☐

c 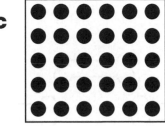   ☐ × ☐ = ☐ × ☐

4 Solve the multiplication problems. Draw arrays in the boxes to help you.

a During PE, the teachers make 10 groups of 4 children.

How many children are there altogether?

☐

b There are 6 tables in the classroom. 4 children sit at each table. How many children are in the classroom?

☐

Now look at and think about each of the *I can* statements.

☐

Date: _____

Name: _____

1 Share the biscuits between 2 bowls.

How many biscuits are in each bowl? ☐

2 Share the biscuits between 4 bowls.

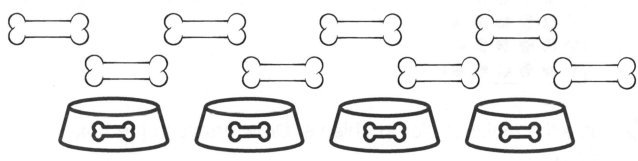

How many biscuits are in each bowl? ☐

3 Share the biscuits between 3 bowls.

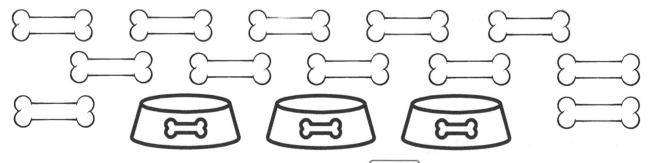

How many biscuits are in each bowl? ☐

4 Share the biscuits between 5 bowls.

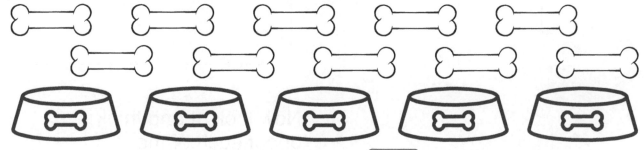

How many biscuits are in each bowl? ☐

5 Solve the divisions. Use the sharing diagrams to help you.

a 8 ÷ 2 = ☐

X X X X X X X X

b 9 ÷ 3 = ☐

X X X X X X X X X

c 20 ÷ 5 = ☐

X X X X X X X X X X X X X X X X X X X X

6 Solve the division problems. Draw sharing diagrams to help you.

a Sophie shares 12 strawberries equally between 4 tarts.

How many strawberries are on each tart? ☐

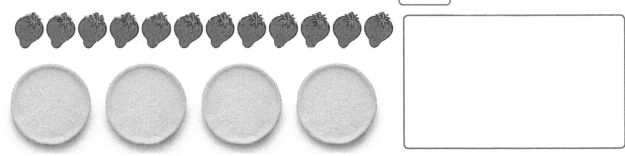

b Share 15 candles equally between 3 cakes.

How many candles are on each cake? ☐

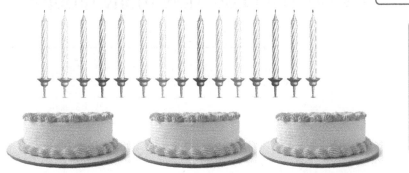

Now look at and think about each of the *I can* statements. ☐

Date: _____

Number

Name: _____

1 Circle groups of 2 to find out how many groups there are.

a How many groups of 2 are in 16? ☐

b How many groups of 2 are in 20? ☐

2 Draw groups of 5 crosses (✗) to find out how many groups of 5 are in 20.

How many groups of 5 are in 20? ☐

3 Draw groups of 10 crosses (✗) to find out how many groups of 10 are in 30.

How many groups of 10 are in 30? ☐

Number

 Solve the divisions. Draw groups to help you.

a $12 \div 2 =$ ☐

b $25 \div 5 =$ ☐

c $40 \div 10 =$ ☐

 Solve the division problems. Draw groups to help you.

a Sunita puts 16 cherries into bags of 8.

How many bags of cherries are there? ☐

b 18 children are going on an outing. 3 children can go in each car.

How many cars are needed? ☐

Now look at and think about each of the *I can* statements. ☐

Date: _____

Name: _____

Number

1 Subtract groups of 3 counters by crossing them out
until you have no counters left. Then count how many groups of
counters you have subtracted.

$15 \div 3 =$ ☐

2 Subtract groups of 4 counters by crossing them out
until you have no counters left. Then count how many groups of
counters you have subtracted.

$24 \div 4 =$ ☐

3 Subtract groups on the number line. Count the jumps
to find the answer.

a $70 \div 10 =$ ☐

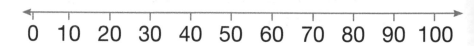

0 10 20 30 40 50 60 70 80 90 100

b $12 \div 2 =$ ☐

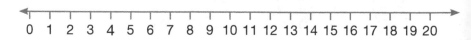

0 1 2 3 4 5 6 7 8 9 10 11 12 13 14 15 16 17 18 19 20

c $45 \div 5 =$ ☐

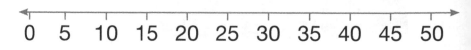

0 5 10 15 20 25 30 35 40 45 50

Solve the divisions. Use the number lines to help you.

a 18 ÷ 2 = ☐

b 40 ÷ 10 = ☐

c 30 ÷ 5 = ☐

 Solve the division problems. Use the pictures to help you.

a Mrs Sim has 20 crayons. She puts 5 crayons on each table.

How many tables are there? ☐

b Mrs Sim has 18 cards and puts 3 cards in each pile.

How many piles does she make? ☐

Now look at and think about each of the *I can* statements. ☐

Date: _____

Number

Number

Name: _____

1 Use the number line to answer the 2 times table facts.

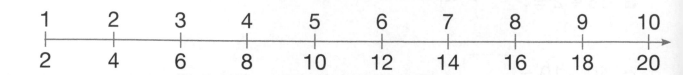

a $2 \times 8 =$ ☐

b $2 \times 5 =$ ☐

c 2×10 ☐

d $2 \times 3 =$ ☐

e $2 \times 6 =$ ☐

f $2 \times 2 =$ ☐

2 Use the number line to answer the 2 times table division facts.

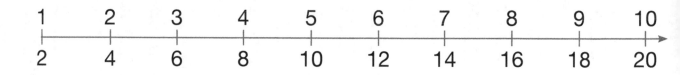

a $10 \div 2 =$ ☐

b $8 \div 2 =$ ☐

c $2 \div 2$ ☐

d $14 \div 2 =$ ☐

e $18 \div 2 =$ ☐

f $20 \div 2 =$ ☐

3 Complete each fact.

a $2 \times 4 =$ ☐

b $2 \times 9 =$ ☐

c $6 \div 2 =$ ☐

d $16 \div 2 =$ ☐

32

Number

4 Use the number line to answer the 5 times table facts.

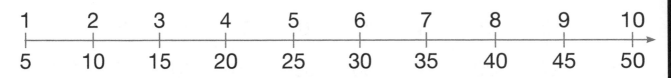

| 1 | 2 | 3 | 4 | 5 | 6 | 7 | 8 | 9 | 10 |
| 5 | 10 | 15 | 20 | 25 | 30 | 35 | 40 | 45 | 50 |

a $5 \times 7 = \boxed{}$

b $5 \times 4 = \boxed{}$

c $5 \times 2 \boxed{}$

d $5 \times 9 = \boxed{}$

e $5 \times 5 = \boxed{}$

f $5 \times 1 = \boxed{}$

5 Use the number line to answer the 5 times table division facts.

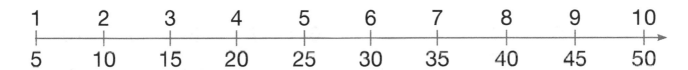

| 1 | 2 | 3 | 4 | 5 | 6 | 7 | 8 | 9 | 10 |
| 5 | 10 | 15 | 20 | 25 | 30 | 35 | 40 | 45 | 50 |

a $50 \div 5 = \boxed{}$

b $35 \div 5 = \boxed{}$

c $25 \div 5 = \boxed{}$

d $40 \div 5 = \boxed{}$

e $15 \div 5 = \boxed{}$

f $30 \div 5 = \boxed{}$

6 Complete each fact.

a $5 \times 3 = \boxed{}$

b $5 \times 8 = \boxed{}$

c $20 \div 5 \boxed{}$

d $45 \div 5 = \boxed{}$

Now look at and think about each of the *I can* statements.

Date: _____

Number

Name: _____

1 Use the number line to answer the 10 times table facts.

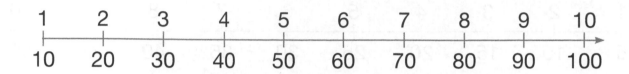

| 1 | 2 | 3 | 4 | 5 | 6 | 7 | 8 | 9 | 10 |
| 10 | 20 | 30 | 40 | 50 | 60 | 70 | 80 | 90 | 100 |

a $10 \times 3 =$ ☐

b $10 \times 5 =$ ☐

c $10 \times 10 =$ ☐

d $10 \times 7 =$ ☐

e $10 \times 2 =$ ☐

f $10 \times 9 =$ ☐

2 Use the number line to answer the 10 times table division facts.

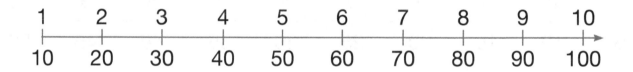

| 1 | 2 | 3 | 4 | 5 | 6 | 7 | 8 | 9 | 10 |
| 10 | 20 | 30 | 40 | 50 | 60 | 70 | 80 | 90 | 100 |

a $10 \div 10 =$ ☐

b $40 \div 10 =$ ☐

c $80 \div 10 =$ ☐

d $60 \div 10 =$ ☐

e $30 \div 10 =$ ☐

f $100 \div 10 =$ ☐

3 Complete each fact.

a $10 \times 4 =$ ☐

b $10 \times 6 =$ ☐

c $20 \div 10 =$ ☐

d $70 \div 10 =$ ☐

Number

 Use the number line to answer the 1 times table multiplication and division facts.

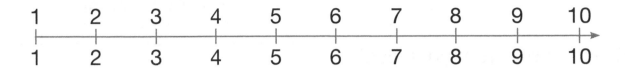

a $1 \times 2 =$ ☐

b $1 \times 10 =$ ☐

c $4 \div 1 =$ ☐

d $6 \div 1 =$ ☐

e $1 \times 8 =$ ☐

f $10 \div 1 =$ ☐

 Use the times table grid to answer the multiplication and division facts.

×	1	2	3	4	5	6	7	8	9	10
1	1	2	3	4	5	6	7	8	9	10
2	2	4	6	8	10	12	14	16	18	20
5	5	10	15	20	25	30	35	40	45	50
10	10	20	30	40	50	60	70	80	90	100

a $5 \times 6 =$ ☐

b $10 \times 8 =$ ☐

c $3 \div 1 =$ ☐

d $10 \div 5 =$ ☐

e $1 \times 5 =$ ☐

f $12 \div 2 =$ ☐

g $50 \div 10 =$ ☐

h $2 \times 7 =$ ☐

Now look at and think about each of the *I can* statements.

☐

Date: _____

Number

Name: _____

1 Draw the **symbol** used in your local currency.

2 Draw a **coin** from your local currency. Make sure you include the currency (if it's shown) and the value of the coin.

3 Draw a **note** from your local currency. Make sure you include the currency (if it's shown) and the value of the note.

4 Draw lines to match the coins and notes to the items with the same value.

 10

 25

 5

 $5

 $1

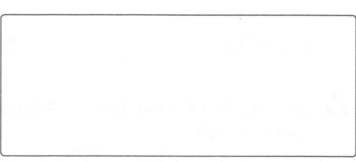 10c

 5c

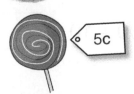 25c

5 In each set, **colour** the coin or note with the **highest** value and **circle** the coin or note with the **lowest** value.

a

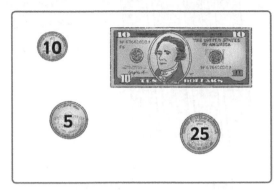

b

c

d

6 Draw lines to match the total in each purse with a coin.

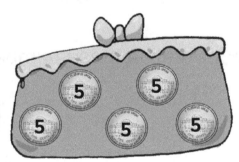

Now look at and think about each of the *I can* statements.

Date: _____

Name: _____

1 How many tens? How many ones?

a

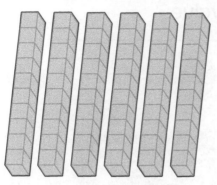

tens: ☐ ones: ☐

b

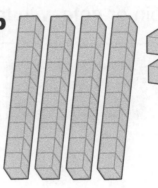 

tens: ☐ ones: ☐

2 How many tens? How many ones?

a

34

tens: ☐ ones: ☐

b

86

tens: ☐ ones: ☐

3 Partition these numbers.

a

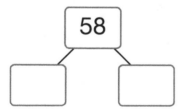

58

b

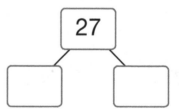

27

4 Fill in the missing numbers.

a

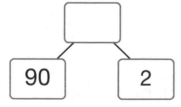

90   2

b

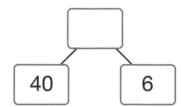

40   6

Circle the **smallest** number in each set.

a 41   75   32   97   **b** 63   26   54   88

c 59   14   43   88   **d** 33   67   35   78

 Circle the **greatest** number in each set.

a 39   73   27   64   **b** 56   48   12   85

c 71   54   90   52   **d** 46   63   47   61

 Circle the 5th crayon and the 12th crayon.

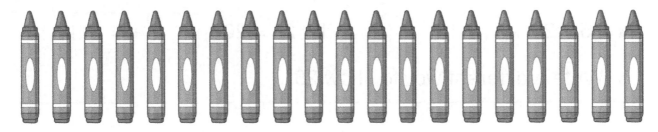

 a Write the position of the  .

b Write the position of the .

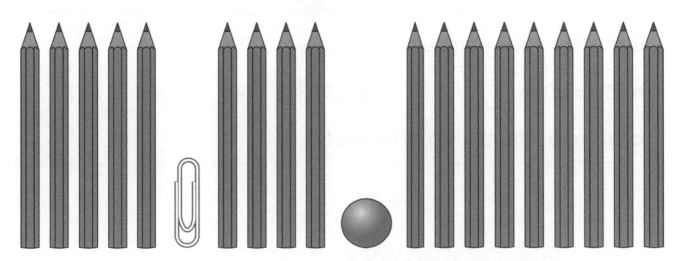

Now look at and think about each of the *I can* statements.

Date: _____

Number

Name: _____

1 Write two number sentences for each number.

a

43

☐ + ☐ = ☐

☐ = ☐ + ☐

b

71

☐ + ☐ = ☐

☐ = ☐ + ☐

2 Complete each number sentence.

a 25 = 20 + ☐

b 25 = 20 + ☐ + ☐

c Write a different way to show 25.

3 Complete each number sentence.

a 37 = 30 + ☐

b 37 = ☐ + ☐ + ☐ + 7

c Write a different way to show 37.

Number

4 Order each set of numbers, starting with the **greatest**.

a 36, 75, 14, 80, 57 ☐ , ☐ , ☐ , ☐ , ☐

b 43, 60, 92, 73, 83 ☐ , ☐ , ☐ , ☐ , ☐

c 75, 36, 11, 38, 46 ☐ , ☐ , ☐ , ☐ , ☐

d 47, 57, 74, 54, 69 ☐ , ☐ , ☐ , ☐ , ☐

5 Write each number on the number line.

a 24

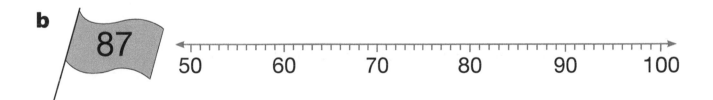

0　　10　　20　　30　　40　　50

b 87

50　　60　　70　　80　　90　　100

6 Round each number to the nearest 10.

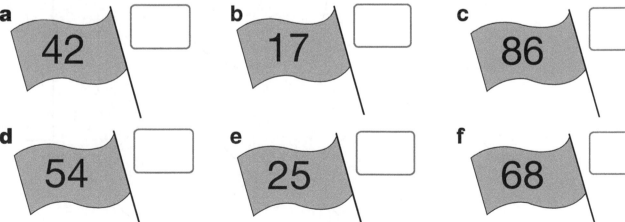

a 42 ☐　　b 17 ☐　　c 86 ☐

d 54 ☐　　e 25 ☐　　f 68 ☐

Now look at and think about each of the *I can* statements. ☐

Date: _____

Number

Name: _____

1 Circle the shapes that have been divided into quarters.

You will need
• coloured pencil

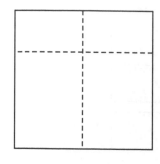
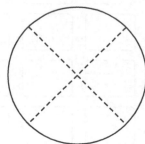
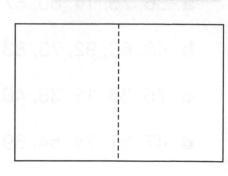

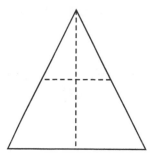
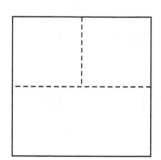
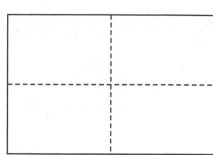

2 Draw lines to divide each shape into quarters.

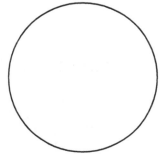

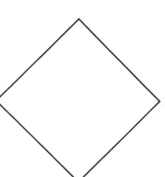

3 Look at the shapes in **2**.

Colour $\frac{1}{4}$ of each shape.

4 Draw dots in each quarter to find $\frac{1}{4}$.

a $\frac{1}{4}$ of 12 = ☐

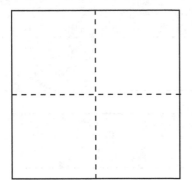

b $\frac{1}{4}$ of 8 = ☐

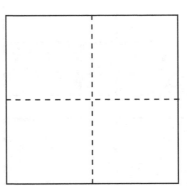

c $\frac{1}{4}$ of 4 = ☐

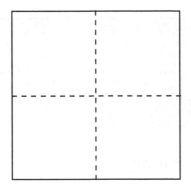

d $\frac{1}{4}$ of 20 = ☐

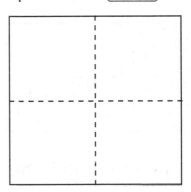

 5 Draw the rest of the counters. Then count how many there are altogether to find how many are in a full set.

a

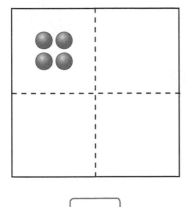

☐

b

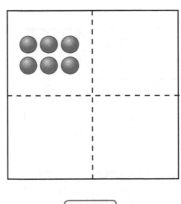

☐

Now look at and think about each of the *I can* statements.

Date: _____

Number

Name: _____

1 Draw lines to match the diagrams that have an equal fraction shaded.

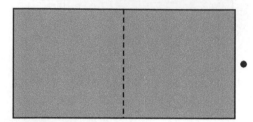
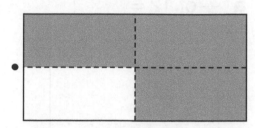

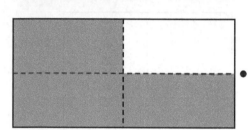
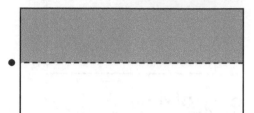

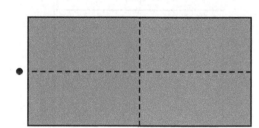

2 Draw lines to match the diagrams to the fractions.

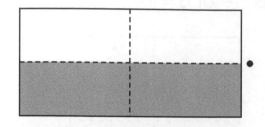   • 1 and $\frac{1}{4}$

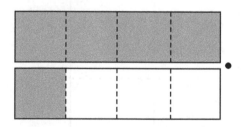   • 1 and $\frac{1}{2}$

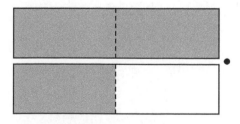   • $\frac{3}{4}$

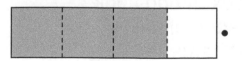   • $\frac{1}{2}$

Number

3 Circle $\frac{1}{2}$ of each set of counters.

a $\frac{1}{2}$ of 12 = ☐

b $\frac{1}{2}$ of 10 = ☐

c $\frac{1}{2}$ of 18 = ☐

d $\frac{1}{2}$ of 14 = ☐

4 Circle $\frac{1}{4}$ of each set of counters.

a $\frac{1}{4}$ of 8 = ☐

b $\frac{1}{4}$ of 12 = ☐

c $\frac{1}{4}$ of 16 = ☐

d $\frac{1}{4}$ of 4 = ☐

5 Complete the number sentences to solve the problems.

a There are 20 children in Blue Class. One quarter of them are 7 years old. How many is this?

☐ ÷ ☐ = ☐

b There are 20 children in Blue Class. Half of them are girls. How many girls are in Blue Class?

☐ ÷ ☐ = ☐

Now look at and think about each of the *I can* statements.

☐

Date: _____

Geometry and Measure

Name: _____

1 Use the calendar to answer the questions.

April

Monday	Tuesday	Wednesday	Thursday	Friday	Saturday	Sunday
		1	2	3	4	5
6	7	8	9	10	11	12
13	14	15	16	17	18	19
20	21	22	23	24	25	26
27	28	29	30			

a How many **days** are there in April? ☐

b What **day** of the week is 17th April? ☐

c How many **Tuesdays** are there in April? ☐

d What day of the week is the **first** day in April? ☐

e What day of the week is the **last** day in April? ☐

2 Number the units of time 1 to 7, starting with the **shortest** (1).

month ☐   week ☐   minute ☐

hour ☐   year ☐   second ☐

day ☐

3 Write the missing months.

a March   April   May   ☐   July

b August   September   October   ☐
December

Write the time each clock shows.

a

b

c

 Show the times on the clocks.

a 10 past 8

b 25 past 2

c 5 to 4

 Write the digital times to match.

a

b

c

___ ___ : ___ ___

___ ___ : ___ ___

___ ___ : ___ ___

Now look at and think about each of the *I can* statements.

Date: _____

Geometry and Measure

Name: _____

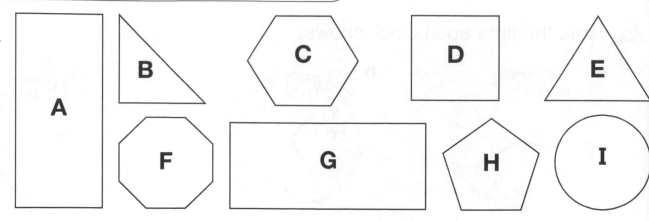

a Sort the shapes by writing the letters in the correct box.

4 sides or fewer		more than 4 sides

b Sort the shapes again by writing the letters in the correct box.

fewer than 4 vertices	4 vertices	more than 4 vertices

2 Use the dot grids to sketch the shapes.

a square   **b** rectangle   **c** triangle

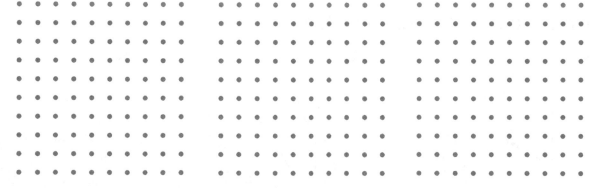

3 Mark the centre of each shape with a cross (**X**).

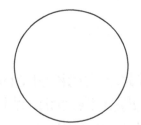

48

Geometry and Measure

4 Circle the shapes and objects that have symmetry.

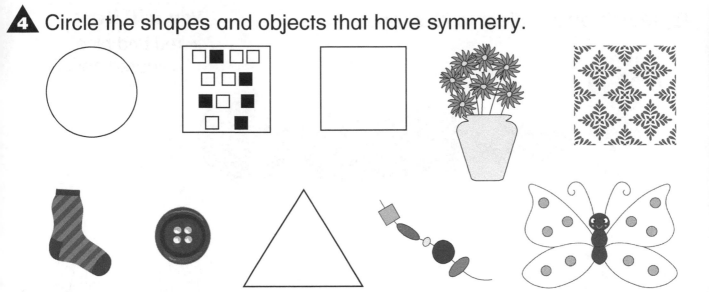

5 Draw lines to show how the letter **E** has turned.

E	⊐	• half turn
E	E	• quarter turn anticlockwise
E	Ш	• full turn
E	Ǝ	• quarter turn clockwise

6 Draw a cross (✗) to show the **right angle** in each shape or object.

Now look at and think about each of the *I can* statements.

Date: _____

Geometry and Measure

Name: _____

1 Match each object to its shape.

You will need
• red and blue coloured pencils

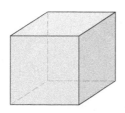

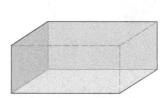

2 Complete the table.

A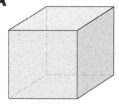
B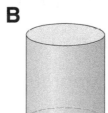
C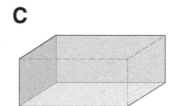
D

	Name of shape	Number of faces	Number of vertices	Number of edges
A				
B				
C				
D				

3 Colour red all the shapes with **5 faces or fewer**.
Colour blue all the shapes with **more than 5 faces**.

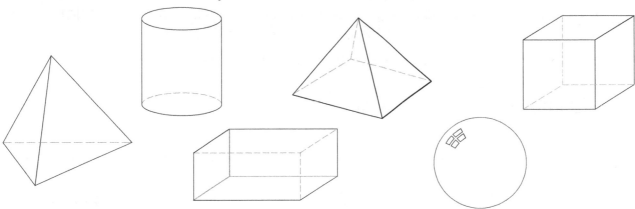

Geometry and Measure

4 Colour red all the shapes with **5 vertices or fewer**.
Colour blue all the shapes with **more than 5 vertices**.

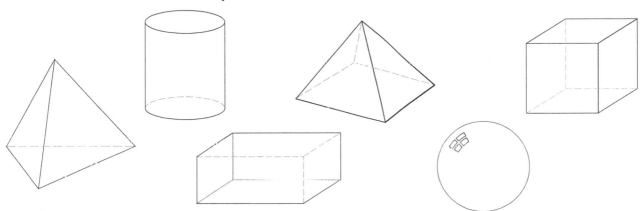

5 Colour red all the shapes with **flat faces** only. Colour blue the shape with a **curved face** only. Circle the shape with **curved** and **flat faces**.

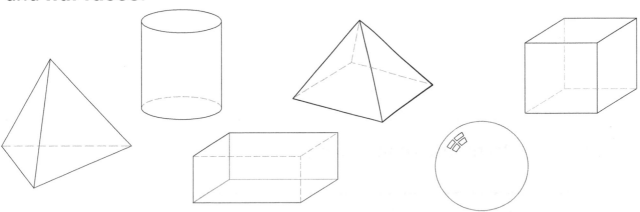

Now look at and think about each of the *I can* statements.

Date: _____

Name: _____

1 Estimate the length of each ribbon using paper clips. Then use the paper clips to measure each ribbon.

You will need
- paper clips
- ruler

a

Estimate: [ ]   Length: [ ]

b

Estimate: [ ]   Length: [ ]

c

Estimate: [ ]   Length: [ ]

2 Write or draw one example of an object in each box.

Shorter than 30 centimetres	Longer than 30 centimetres	Longer than 1 metre

3 Use a ruler to measure these lines.

a ——————————————— [ ]

b ————————————————————

[ ]

4 Use a ruler to draw a line 7 centimetres long.

5 Use a ruler to draw a line 11 centimetres long.

6 Estimate the height of each ribbon, then use a ruler to measure it.

a   b

Estimate: ☐

Height: ☐

Estimate: ☐

Height: ☐

7 Use a ruler to draw a line 9 cm tall.

Now look at and think about each of the *I can* statements.

☐

Date: _____

Name:
_____

1 Circle the measurement you would use to find the mass of each object.

a

grams
kilograms

b

grams
kilograms

c

grams
kilograms

2 What is the mass of each piece of fruit?

a
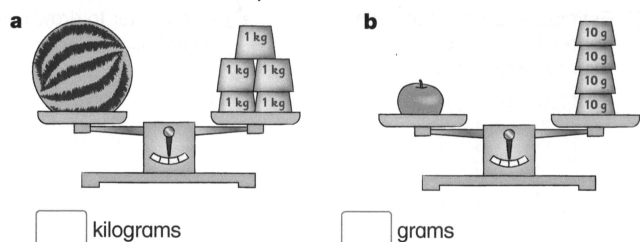

1 kg

1 kg 1 kg

1 kg 1 kg

10 g

10 g

10 g

10 g

b

[ ] kilograms

[ ] grams

3 What is the mass of each bag of food?

a
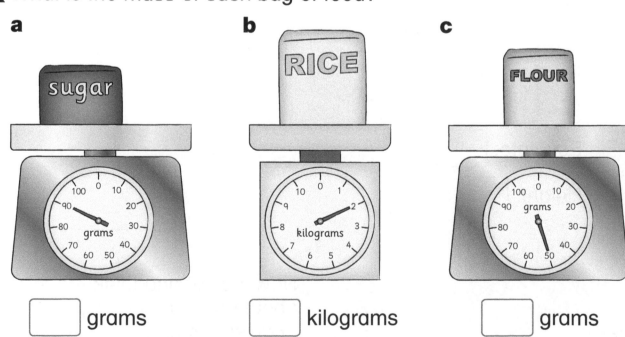

sugar

100 0 10
90   20
80   30
grams
70   40
60 50

RICE

10 0 1
9   2
8   3
kilograms
7   4
6 5

FLOUR

100 0 10
90 grams 20
80   30
70   40
60 50

[ ] grams

[ ] kilograms

[ ] grams

4 Show the mass on the scales.

a 9 kilograms   **b** 5 kilograms   **c** 80 grams

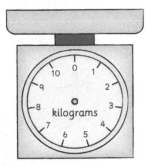 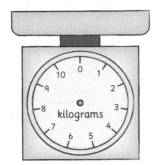 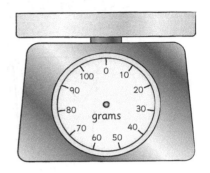

Geometry and Measure

5 Find the mass of each box. Then complete the sentence.

a   **b**   **c**

☐ kilograms   ☐ kilograms   ☐ kilograms

d Box C is ☐ kilograms **heavier** than Box B.

6 Find the mass of each sack. Then complete the sentence.

a   **b**   **c**

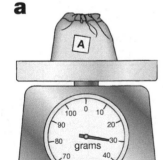 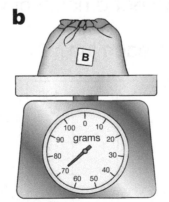 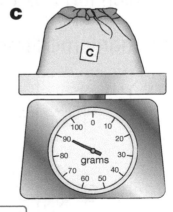

☐ grams   ☐ grams   ☐ grams

d Sack ☐ is **lighter** than Sack B.

Now look at and think about each of the *I can* statements. ☐

Date: _____

Name: _____

1 Look at each pair of containers. Circle the container that holds **more**.

a
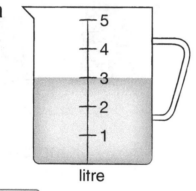

b

c

2 How many litres are in each jug?

a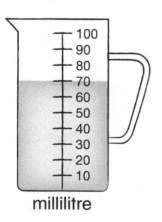
litre

☐ litres

b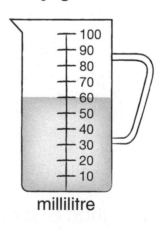
litre

☐ litres

c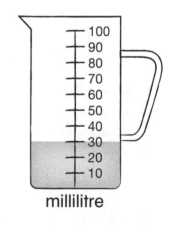
litre

☐ litres

d Which jug has the **smallest** amount of water? ☐

3 How many millilitres are in each jug?

a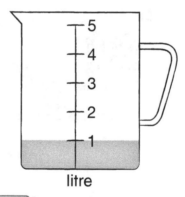
millilitre

☐ millilitres

b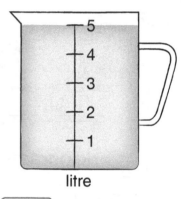
millilitre

☐ millilitres

c
millilitre

☐ millilitres

d Which jug has the **most** water? ☐

4 Circle the objects that would hold **less** than 1 litre.

5 Colour each jug to show how many litres are in it.

a 4 litres

litre

b 3 litres

litre

c 2 litres

litre

6 Colour each jug to show many millilitres are in it.

a 90 millilitres

millilitre

b 40 millilitres

millilitre

c 20 millilitres

millilitre

7 Write the temperature shown on each thermometer.

a

b

c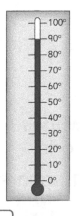

d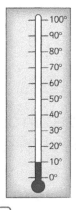

☐ degrees ☐ degrees ☐ degrees ☐ degrees

Now look at and think about each of the *I can* statements.

Date: _____

Geometry and Measure

Name: _____

Geometry and Measure

1 Sketch the reflection of each shape. Use a mirror to help you.

You will need
- mirror   - red pencil

a

b

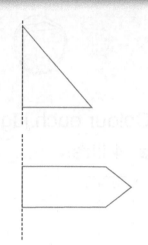

c

d

2 Use the words in the box to complete the sentences.

| forwards   backwards   left   right |

a The  moves 2 squares

[_____].

b The 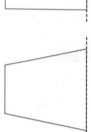 moves 3 squares

[_____].

c The 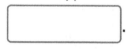 moves 2 squares

[_____].

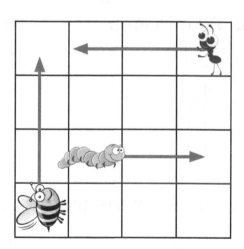

3 Use arrows to show these movements on the grid.

a The moves 1 square right.

b The moves 3 squares backwards.

c The moves 2 squares left.

4 Use the words in the box to complete the sentences.

| above |
| below |
| to the right of |
| to the left of |

a The is [_____] the .

b The is [_____] the .

c The is [_____] the .

d The is [_____] the  .

5 The arrow shows a half turn clockwise.

On each circle, use a red pencil to draw an arrow showing the direction.

a a quarter turn clockwise

b a whole turn anticlockwise

c a half turn anticlockwise

Now look at and think about each of the *I can* statements.

Date: _____

Name: _____

1 Complete the tally chart.

You will need
• coloured pencil

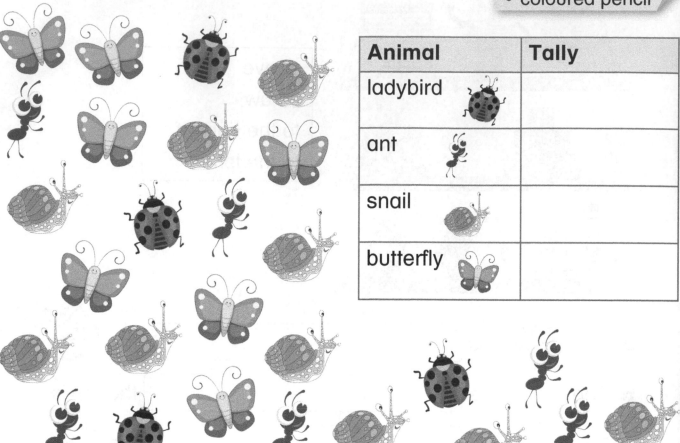

Animal	Tally
ladybird	
ant	
snail	
butterfly	

2 Use the data in the tally chart in **1** to draw a pictogram to show how many there are of each animal.

⚘ = 1 animal

Animals

Animal	Number
ladybird	
ant	
snail	
butterfly	

3 Use the data in the tally chart in **1** to draw a block graph to show how many there are of each animal.

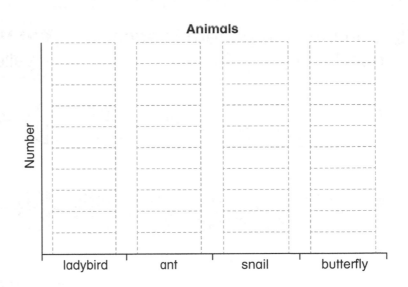

Animals

Number

ladybird   ant   snail   butterfly

4 Answer these questions.

a Which animal is there the most of?

b How many more snails are there than butterflies?

c How many ants and ladybirds are there altogether?

5 Complete the Carroll diagram.

ladybird  ant

snail butterfly

	has wings	does not have wings
has spots		
does not have spots		

6 Look at the animals in **1**. What is the mistake in this Venn diagram?

_____

_____

_____

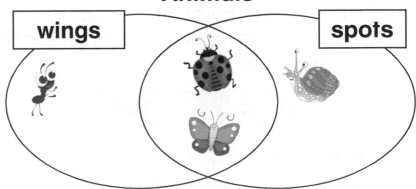

Animals

wings   **spots**

Now look at and think about each of the *I can* statements.

Date: _____

Statistics and Probability

Name: _____

1 Continue the pattern in each bead string.

You will need
• yellow and red coloured pencils

a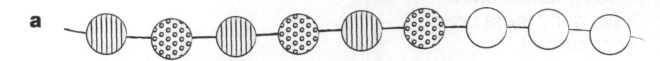

b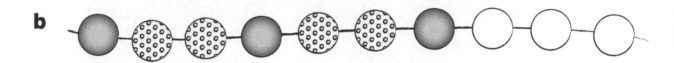

c

2 Colour the beans to make it **certain** you would pick a red bean.

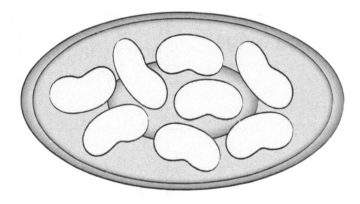

3 Colour the beans so that you could pick **either** a red or a yellow bean.

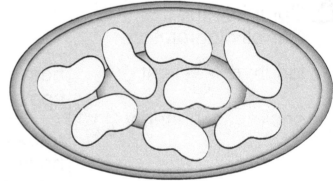

4 Class 2 investigated the chance of getting heads when a coin is flipped. Is the result in this tally chart possible? Explain.

	Tally	Total
heads	//// //// ////	14
tails	//// //// ///	13

_____

_____

_____

Statistics and Probability

5 This pictogram shows the beans chosen by 15 children in Class 2.

 = 1 child

Beans

Colour	Number
red	🫘 🫘 🫘 🫘 🫘 🫘 🫘 🫘 🫘
yellow	🫘 🫘 🫘 🫘 🫘 🫘

a Which colour bean is more popular?

b How many children chose this colour?

c How many more children chose this colour than the other colour?

6 Use the data in the pictogram in **5** to draw a block graph to show the beans chosen by the children in Class 2.

Beans

Number of beans

red   yellow

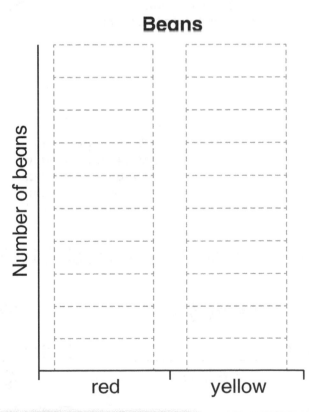

Now look at and think about each of the *I can* statements.

Date: _____

The Thinking and Working Mathematically Star

Think about each of these *I can* statements and record how confident you feel about **Thinking and Working Mathematically**.

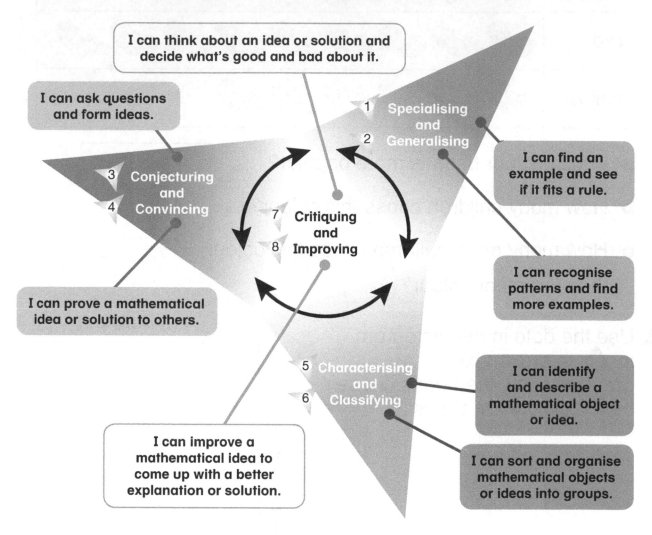

I can think about an idea or solution and decide what's good and bad about it.

I can ask questions and form ideas.

1
2
Specialising and Generalising

I can find an example and see if it fits a rule.

3
Conjecturing and Convincing
4

7
Critiquing and
8
Improving

I can recognise patterns and find more examples.

I can prove a mathematical idea or solution to others.

5
Characterising and
6
Classifying

I can identify and describe a mathematical object or idea.

I can improve a mathematical idea to come up with a better explanation or solution.

I can sort and organise mathematical objects or ideas into groups.

How confident do you feel Thinking and Working Mathematically?

Term ❶	Term ❷	Term ❸
Date: _____	Date: _____	Date: _____
☺ ☺ ☹	☺ ☺ ☹	☺ ☺ ☹
Date: _____	Date: _____	Date: _____
☺ ☺ ☹	☺ ☺ ☹	☺ ☺ ☹